Using Simple Machines

Wheels and Axles All Around

by Trudy Becker

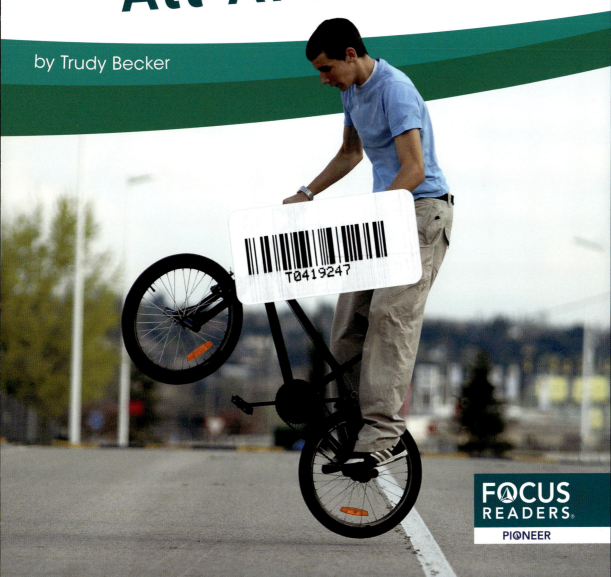

FOCUS READERS
PIONEER

www.focusreaders.com

Copyright © 2024 by Focus Readers®, Lake Elmo, MN 55042. All rights reserved. No part of this book may be reproduced or utilized in any form or by any means without written permission from the publisher.

Focus Readers is distributed by North Star Editions:
sales@northstareditions.com | 888-417-0195

Produced for Focus Readers by Red Line Editorial.

Photographs ©: iStockphoto, cover, 1, 4, 6, 8; Shutterstock Images, 10, 12 (top), 12 (bottom), 14, 17, 18, 20

Library of Congress Cataloging-in-Publication Data
Names: Becker, Trudy, author.
Title: Wheels and axles all around / by Trudy Becker.
Description: Lake Elmo, MN : Focus Readers, [2024] | Series: Using simple
   machines | "Focus Readers Pioneer"-- cover. | Includes bibliographical
   references and index. | Audience: Grades K-1
Identifiers: LCCN 2023003208 (print) | LCCN 2023003209 (ebook) | ISBN
   9781637396025 (hardcover) | ISBN 9781637396599 (paperback) | ISBN
   9781637397718 (ebook pdf) | ISBN 9781637397169 (hosted ebook)
Subjects: LCSH: Wheels--Juvenile literature. | Axles--Juvenile literature.
Classification: LCC TJ181.5 .B43 2024 (print) | LCC TJ181.5 (ebook) | DDC
   621.8--dc23/eng/20230208
LC record available at https://lccn.loc.gov/2023003208
LC ebook record available at https://lccn.loc.gov/2023003209

Printed in the United States of America
Mankato, MN
082023

About the Author

Trudy Becker lives in Minneapolis, Minnesota. She likes exploring new places and loves anything involving books.

Table of Contents

CHAPTER 1
Roll the Weight 5

CHAPTER 2
What Are Wheels and Axles? 9

CHAPTER 3
Wheels and Axles Everywhere 13

THAT'S AMAZING!
Making Power 16

CHAPTER 4
Fun with Wheels and Axles 19

Focus on Wheels and Axles • 22
Glossary • 23
To Learn More • 24
Index • 24

Chapter 1

Roll the Weight

A girl needs to carry some tools across her yard. The tools are heavy. So, the girl puts them in a wheelbarrow. She pushes the wheelbarrow. Now the job is easier.

The girl's wheelbarrow has a wheel. It has an **axle**, too. The axle helps the wheel roll. That lets her move the weight more easily. A wheel and axle is one of the six **simple machines**.

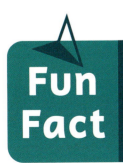

Fun Fact

Some machines have one wheel. Some have many wheels.

Chapter 2

What Are Wheels and Axles?

All simple machines help people do jobs. People use wheels and axles to move things. It can be faster and easier than other ways to move things.

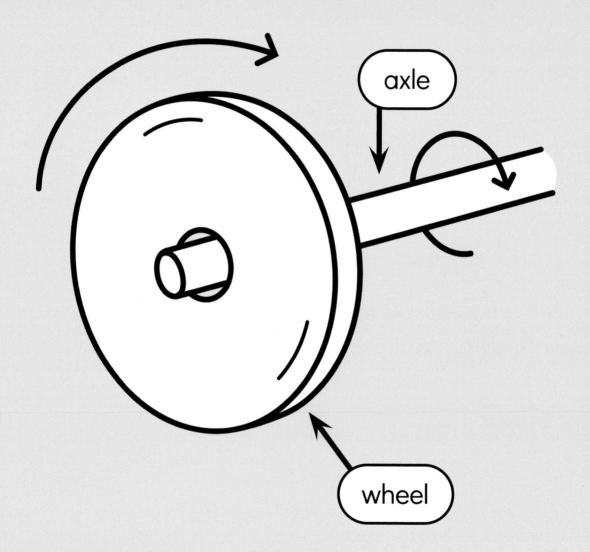

A wheel and axle has two parts. The first is the wheel. It has a circle shape. The other part is the axle. This bar is attached to the middle of the wheel. It helps the wheel spin. Wheels and axles work by rolling.

Fun Fact In some machines, the wheel and axle both move. In others, only the wheel moves.

Chapter 3

Wheels and Axles Everywhere

Wheels and axles are all around. Many kinds of **transportation** use them. Cars and buses have wheels. Planes use wheels to take off and land. Trains have many wheels, too.

Wheels and axles are also in people's homes. Some drawers open with wheels on tracks. A pizza cutter has a wheel. People add **force** using the handle. The wheel turns.

Fun Fact

A pizza cutter also uses a **wedge**. The wedge's sharp edge cuts the pizza.

That's Amazing!

Making Power

Wheels and axles can help make power. Some people use **waterwheels**. First, water flows onto a wheel. The weight makes the wheel and axle move. That movement powers a machine. Waterwheels are not common today. But wheels are still part of many **modern** machines.

Chapter 4

Fun with Wheels and Axles

Wheels and axles are helpful for jobs. But they are used for fun, too. Ferris wheels are wheels and axles. People ride in them. The giant wheel turns. People move higher and lower.

Many people use bicycles for fun. Riders add force to the pedals. The pedals connect to the axles. That makes the wheels turn. People can ride all around.

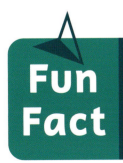

Fun Fact Roller skates and scooters are fun ways to use wheels.

FOCUS ON
Wheels and Axles

Write your answers on a separate piece of paper.

1. Write a sentence that explains the main idea of Chapter 2.

2. What is the most helpful wheel and axle that you use? Why?

3. What other simple machine is part of a pizza cutter's blade?
   - A. a handle
   - B. a wedge
   - C. a pedal

4. How can wheels help move a heavy weight?
   - A. Wheels let the weight roll.
   - B. Wheels push the weight deep into the ground.
   - C. Wheels let the weight travel up and down.

Answer key on page 24.

Glossary

axle
A bar that a wheel spins on.

force
A push or pull that changes how something moves.

modern
Using new and improved ideas and tools.

simple machines
Machines with only a few parts that make work easier.

transportation
Ways of moving people and goods from one place to another.

waterwheels
Large wheels moved by flowing water. They help machines work.

wedge
A triangular shape that can cut or hold something. It is one of the six simple machines.

To Learn More

BOOKS

Blevins, Wiley. *Let's Find Wheels and Axles*. North Mankato, MN: Capstone Press, 2021.

Mattern, Joanne. *Wheels and Axles*. Minneapolis: Bellwether Media, 2020.

NOTE TO EDUCATORS

Visit **www.focusreaders.com** to find lesson plans, activities, links, and other resources related to this title.

Index

B
bicycles, 21

R
rolling, 7, 11

T
transportation, 13, 21

W
wheelbarrow, 5, 7

Answer Key: **1.** Answers will vary; **2.** Answers will vary; **3.** B; **4.** A